Y0-DBZ-382

WITHDRAWN

FIELD TRIPS
FARMS

Kathleen Reitmann and Heather Kissock

AV2

www.av2books.com

Step 1
Go to **www.av2books.com**

Step 2
Enter this unique code

YHMTDGOQ6

Step 3
Explore your interactive eBook!

AV2 is optimized for use on any device

Your interactive eBook comes with...

Contents
Browse a live contents page to easily navigate through resources

Audio
Listen to sections of the book read aloud

Videos
Watch informative video clips

Weblinks
Gain additional information for research

Try This!
Complete activities and hands-on experiments

Key Words
Study vocabulary, and complete a matching word activity

Quizzes
Test your knowledge

Slideshows
View images and captions

... and much, much more!

FIELD TRIPS
FARMS

Contents

- 2   AV2 Book Code
- 4   Why Visit a Farm?
- 6   Down on the Farm
- 8   Farms around the World
- 10   The Farmstead
- 12   What to See on a Farm
- 14   Things to Do on a Farm
- 16   Farm Rules
- 18   A History of Farms
- 20   Who Works on a Farm?
- 22   10 Questions to Test Your Knowledge
- 23   Key Words/Index

Why Visit a Farm?

Most of the food people eat comes from some kind of farm. Animals such as cattle and pigs are raised on farms. Wheat, barley, and other **crops** are grown on farms. These grains are used to make bread and other foods.

Paying a visit to a farm is a way to find out more about where food comes from. People can also learn about the role farmers play in putting food on tables across the country. By visiting a farm, people can see the important work that farmers do.

Visiting a farm can give people the chance to learn about raising a variety of animals, including chickens.

The **United States** is home to more than **2 million farms**.

Approximately **30%** of American farmers are **women**.

About **99%** of U.S. farms are **family owned**.

Down on the Farm

Most farms focus on one area of production. Some farmers may raise a specific type of **livestock**. Others may grow one or more kinds of crops. There are also farms that have both livestock and crops.

Farms can be grouped in many different ways. For example, they can be family owned or **commercial**. The most common way to group farms, however, is by the products raised or grown on them.

Dairy Farms
Dairy farms produce milk and milk products, such as butter, cheese, ice cream, or yogurt. Many dairy farms rely on cows for milk. However, on some dairy farms, goats or sheep may be raised instead.

Livestock Farms
Livestock farms include cattle ranches and pig farms. The animals on these farms are raised solely to be made into food.

Crop Farms

Crop farms are often known as places where grains are grown. However, fruits and vegetables are also considered crops. A farm that specializes in fruit is sometimes called an orchard.

Mixed Farms

A mixed farm has both crops and livestock. These farms allow farmers to have more than one source of **income**. They are often smaller in size than other types of farms.

Poultry Farms

Poultry farms are places where birds are raised for their meat, eggs, and feathers. Chickens, turkeys, ducks, and geese are the main types of birds raised on poultry farms.

Fish Farms

Today, about half of the fish sold as food have been raised on fish farms. The fish are kept in water in enclosures either outside or in buildings. Fish commonly farmed include salmon, halibut, and tuna.

Farms around the World

Farming is an important industry for most countries. The type of farming varies, depending on location and the resources available.

1
King Ranch
Houston, United States

King Ranch is often called the "Birthplace of American Ranching." Founded in 1853, it is the largest ranch in the United States. At 825,000 acres (333,866 hectares), it is larger than the state of Rhode Island.

2
Bom Futuro
Cuiabá, Brazil

Cuiabá, Brazil, serves as the headquarters for Bom Futuro, the world's largest producer of soybeans. The corporation's farmland covers an area of about 1.2 million acres (486,000 ha).

Field Trips

3 Porte de Versailles
Paris, France

In 2019, efforts were underway to build the world's largest rooftop farm in Paris. The farm will cover 150,695 square feet (14,000 sq. meters). More than 30 varieties of plants will be grown there.

4 Limmu Coffee Farm
Jimmu, Ethiopia

Ethiopia is the fifth-largest coffee producer in the world. The Limmu Coffee Farm is the country's largest modern coffee **plantation**. It covers an area of more than 29,653 acres (12,000 ha).

5 Shenlan 1
Yellow Sea, China

In 2018, China began operating its first deep-sea fish farm, with the construction of its Shenlan 1 salmon cage. Shenlan 1 is the world's first fully **submersible** fish cage. It can hold up to 300,000 salmon.

6 Anna Creek Station
William Creek, Australia

At 3.9 million acres (1.6 million ha), Anna Creek Station, in South Australia, is the world's largest cattle farm. This family-owned ranch is home to more than 9,500 cattle.

Silos

Silos hold the grain **harvested** from a farm's crops. Some silos are **insulated** and temperature controlled. This makes sure the grain does not spoil.

Machinery Buildings

Machinery buildings store farm equipment that is not in use. The buildings must be large and open so machinery can be easily moved in and out.

Livestock Barns

Barns are places where livestock can rest, eat, drink, and be milked. Some barns have **stalls** to separate the animals. Other barns have an open floor that allows livestock to roam freely.

The Farmstead

When people think of farms, they often picture specific types of buildings, such as barns. This is because most farms are engaged in **agriculture**. Whether the farms have livestock, crops, or both, they need storage buildings, living quarters, and machinery sheds. Most farms also have open land. This is used for growing crops or as **pasture**. Not all farms look the same, however. The placement of buildings depends on several factors. These include climate, soil conditions, and access to roads.

Farmhouses

The farmhouse is the center of a family-owned farm. It is where the family lives and often serves as the farm's business office.

Grain farms usually have good access to railway lines. This is because farmers rely on trains to take the grain to market. The grain is typically stored in elevators beside the track.

What to See on a Farm

People have farmed for thousands of years. Over time, farmers have found new ways to get their work done. A visit to a farm can show visitors how farming used to be done, how it is done today, and what the future of farming might look like.

Old Windmill Farm

Visitors to this farm near Lancaster, Pennsylvania, learn about the traditional farming methods used by **Amish** people. Guests can hand-milk cows, make butter using a **churn**, and gather eggs from the chicken coop.

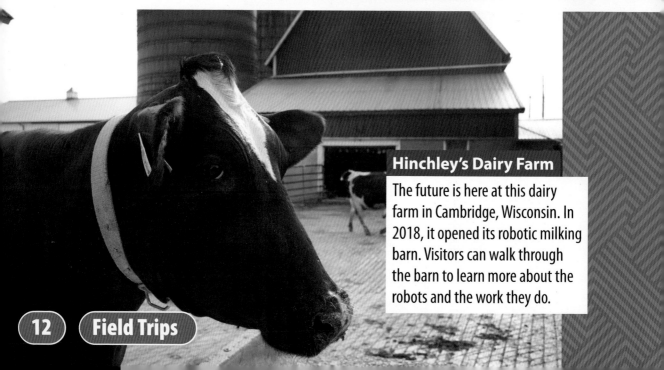

Hinchley's Dairy Farm

The future is here at this dairy farm in Cambridge, Wisconsin. In 2018, it opened its robotic milking barn. Visitors can walk through the barn to learn more about the robots and the work they do.

James Ranch
The focus of this farm in Durango, Colorado, is on **organic** farming. Cattle feed only on plants. Eggs come from **free-range** hens. Vegetable crops are grown without the use of chemicals.

VertiCulture Farms
This rooftop farm in Brooklyn, New York, is both a crop farm and a fish farm. The crops and fish help each other grow. The plants **filter** water for the fish, while the fish provide food for the plants. This form of farming is called aquaponics.

Visitors to **Old Windmill Farm** can tour a barn that was **built** in the **1700s**.

The **robots** at Hinchley's Dairy Farm **milk** about **240 cows** a day.

VertiCulture Farms produces almost **60,000 pounds** (27,216 kilograms) of **fresh herbs** and **greens per year**.

Farms 13

Tour guides can provide visitors with unique experiences. These can include visiting a chicken coop to learn more about egg laying.

14 Field Trips

Things to Do on a Farm

Many working farms offer tours to school groups and the general public. Most tours are led by a guide. This person knows all about the farm and the work that goes into taking care of it. As the group is taken through the farm, the guide shares this knowledge with everyone.

Sometimes, farms hold special activities for visitors. The farms may offer wagon rides through the crop fields. There may be a petting zoo full of farm animals to cuddle. Some farms hold classes as well. Guests can learn how to make cheese or homemade bread.

Petting zoos let visitors handle various farm animals and care for them for a short time.

Farms

Farm Rules

Visiting a working farm is a fun learning adventure. For the safety of the people and the animals, it is important to be respectful and follow the rules of the farm.

6 Simple Rules

1. Stay with your group. Do not wander off on your own.
2. Listen to the farmer when he or she speaks.
3. Do not run or make loud noises. You could scare the animals.
4. Do not touch an animal unless you have permission.
5. Wash your hands after handling animals. You do not want to spread germs.
6. Do not climb on any fences or machinery.

It is important to ask for permission before feeding farm animals. Some animals may be on special diets.

A History of Farms

Farming has undergone many changes over the years. In the past, most work was done by hand or with animals. Today, technology has led to increased production.

Explorer Christopher Columbus's arrival in North America launches the Columbian Exchange, a period that sees the exchange of crops and animals between Europe, Africa, and the Americas.

12,000 BC

5000 BC

1492 AD

Large-scale farming begins in Egypt.

Wild cereal grains are grown in Southwest Asia and North Africa.

English farmer Jethro Tull creates a mechanical drill that plants seeds in a row.

A sheep named Dolly is born. She is the first **cloned** farm animal.

1701 **1892** **1996** **2019**

The U.S. government announces that it will give farmers $16 billion in aid to help them survive a poor economy. This is in addition to $12 billion given to farmers the year before.

John Froelich invents the first gas-powered tractor. This machine eventually takes over the farmwork that horses once did.

Farms 19

Who Works on a Farm?

Working farms require the help of many people. Some operate machines that harvest crops. Others take care of livestock. Most farm jobs take place outdoors.

Farm and Ranch Workers

These people are responsible for caring for the animals on a farm or ranch. They make sure the animals are fed on time and that their housing is kept clean.

Animal Breeders

A breeder's job is to find animals that represent the best of the **species**. They then breed these animals with the hope that they will produce high-quality offspring.

Agricultural Equipment Operators

Farming uses a variety of machinery, ranging from harvesters to plows. Equipment operators know how to use the machinery in a safe and effective manner.

Crop Laborers

These workers help take care of the plants growing in orchards and on other crop farms. Besides planting crops, laborers are also responsible for watering them and for keeping them free of insects and other pests.

Farms

10 Questions to Test Your Knowledge

1. How many farms does the United States have?
2. What is a mixed farm?
3. Where is the world's largest cattle farm?
4. What type of building holds the grain harvested from a farm's crops?
5. What factors determine the placement of buildings on a farm?
6. What does Hinchley's Dairy Farm use to help milk its cows?
7. What is an organic farm?
8. Why is it important to wash your hands after touching an animal?
9. Who invented the first gas-powered tractor?
10. What do animal breeders do on a farm?

ANSWERS: 1. More than 2 million **2.** A farm with both crops and livestock **3.** South Australia, Australia **4.** Silo **5.** Soil conditions, access to roads, and climate **6.** Robots **7.** A farm that grows crops using natural fertilizers and pesticides **8.** To not spread germs **9.** John Froelich **10.** Find animals that represent the best of the species and breed them

Key Words

agriculture: the science or practice of cultivating the soil, producing crops, and raising livestock

Amish: members of a religious group who live in a simple, traditional way

churn: a container in which butter is made

cloned: made to be identical to another organism

commercial: intended to make a profit

crops: cultivated plants that are grown as food

filter: to remove unwanted materials

free-range: kept in natural conditions, with freedom of movement

harvested: gathered ripened crops

income: money received for providing a good or service

insulated: protected by something that stops heat from escaping or entering

livestock: animals kept or raised on a farm

organic: a method of growing crops that uses natural fertilizers and pesticides

pasture: grassy land suitable for grazing animals

plantation: a large farm dedicated to planting crops on a large scale

species: a group of animals that shares common traits

stalls: separate compartments for animals in a barn

submersible: designed to operate underwater

Index

barn 10, 11, 12, 13
Bom Futuro 8

crop farms 7, 13, 21

dairy farms 6, 12, 13, 22
Dolly 19

Egypt 18

fish farms 7, 9, 13
Froelich, John 19, 22

harvest 10, 20, 22
Hinchley's Dairy Farm 12, 13, 22

James Ranch 13

King Ranch 8

livestock farms 6

mixed farms 7, 22

Old Windmill Farm 12, 13

poultry farms 7

silos 10, 22

tours 13, 15
tractor 19, 22

VertiCulture Farms 13

Farms 23

Get the best of both worlds.
AV2 bridges the gap between print and digital.

The expandable resources toolbar enables quick access to content including **videos**, **audio**, **activities**, **weblinks**, **slideshows**, **quizzes**, and **key words**.

Animated videos make static images come alive.

Resource icons on each page help readers to further **explore key concepts**.

Published by AV2
14 Penn Plaza 9th Floor
New York, NY 10122
Website: www.av2books.com

Copyright © 2021 AV2
All rights reserved. No part of this publication may be reproduced, stored in a retrieval system, or transmitted in any form or by any means, electronic, mechanical, photocopying, recording, or otherwise, without the prior written permission of the publisher.

Library of Congress Cataloging-in-Publication Data
Names: Reitmann, Kathleen, author. | Kissock, Heather, author.
Title: Farms / Kathleen Reitmann and Heather Kissock.
Description: New York, NY : AV2, 2021. | Series: Field trips | Includes index. | Audience: Grades 4-6
Identifiers: LCCN 2020010958 (print) | LCCN 2020010959 (ebook) | ISBN 9781791121556 (library binding) | ISBN 9781791121563 (paperback) | ISBN 9781791121570 | ISBN 9781791121587
Subjects: LCSH: Farms--Juvenile literature. | Farm life--Juvenile literature.
Classification: LCC S519 .R45 2021 (print) | LCC S519 (ebook) | DDC 630--dc23
LC record available at https://lccn.loc.gov/2020010958
LC ebook record available at https://lccn.loc.gov/2020010959

Printed in Guangzhou, China
1 2 3 4 5 6 7 8 9 0 24 23 22 21 20

052020
101119

Editor: Heather Kissock
Designer: Terry Paulhus

Every reasonable effort has been made to trace ownership and to obtain permission to reprint copyright material. The publishers would be pleased to have any errors or omissions brought to their attention so that they may be corrected in subsequent printings.

AV2 acknowledges Getty Images, Alamy, iStock, Shutterstock, and Old Windmill Farm as its primary image suppliers for this title.